# KAVA:
## Nature's Answer to Anxiety

by Ray Sahelian, M.D.
Author of the best-selling *Melatonin, DHEA, Creatine,
St. John's wort, Glucosamine,* and *Pregnenolone,* and editor
of *Longevity Research Update*

This booklet is intended for educational and informational purposes only. Please see a qualified healthcare professional if you have questions about your health.

Copyright © 1997 Longevity Research Center, Inc.

All rights reserved. No part of this book may be reproduced, stored in an electronic retrieval system, or transcribed in any form or by any means, electronic or mechanical, including photocopying and recording, without the prior written permission of the author except for the inclusion of quotations in a review.

Published by:
IMPAKT Communications, Inc.
P.O. Box 12496
Green Bay, WI 54307-2496
(920) 434-3838 (U.S.)
(604) 421-5887 (Canada)

# Contents

Tales of the South Pacific . . . . . . . . . . . . . . . . . . . . . . . . . . 5
What is kava? . . . . . . . . . . . . . . . . . . . . . . . . . . . . . . . . . . . 7
What's in kava? . . . . . . . . . . . . . . . . . . . . . . . . . . . . . . . . . 7
How will kava make me feel? . . . . . . . . . . . . . . . . . . . . . . . 8
A brief history . . . . . . . . . . . . . . . . . . . . . . . . . . . . . . . . . . 9
The study that convinced me . . . . . . . . . . . . . . . . . . . . . . 10
For what conditions is kava useful? . . . . . . . . . . . . . . . . . 11
Other properties of kava . . . . . . . . . . . . . . . . . . . . . . . . . 14
How does it work? . . . . . . . . . . . . . . . . . . . . . . . . . . . . . 15
What's the right dosage? . . . . . . . . . . . . . . . . . . . . . . . . . 15
What you will find in your vitamin or retail store . . . . . . . . . 16
How long can I take kava? . . . . . . . . . . . . . . . . . . . . . . . 17
What about side effects? . . . . . . . . . . . . . . . . . . . . . . . . 17
Caution . . . . . . . . . . . . . . . . . . . . . . . . . . . . . . . . . . . . . 18
Kava compared to pharmaceutical drugs . . . . . . . . . . . . . . 19
Can I switch from tranquilizers to kava? . . . . . . . . . . . . . . 20
Can I combine tranquilizers with kava? . . . . . . . . . . . . . . 20
Kava and alcohol . . . . . . . . . . . . . . . . . . . . . . . . . . . . . . 21
Combining kava with nutrients and hormones . . . . . . . . . . 21
Interviews with kava experts . . . . . . . . . . . . . . . . . . . . . . 22
Summary and practical recommendations . . . . . . . . . . . . . 25

# Kava: Nature's Answer to Anxiety

## Tales of the South Pacific

"When we sip kava, we forget there's a class system. There are no longer Ph.D.s, princes, preachers, nor paupers. Everyone opens up, sings, and dances together."

This was my introduction to a kava ceremony at the National Nutritional Foods Association convention in Las Vegas, Nevada, in July 1997. One of the companies marketing a kava product had invited natives of the country Tonga (a small island in the South Pacific) to share with the attendees an actual kava ceremony. A group of curious vitamin store managers were seated in a circle on a wide piece of green artificial turf, facing a group of six Tongans in native costumes. The set was decorated with small palm trees and tropical flowers, providing a Pacific island-like atmosphere. A large wooden bowl filled with a brownish liquid was placed on the green turf. A middle-aged Tongan woman with a serene smile was gently stirring this liquid with a wooden spatula. Next to her was the emcee of the ceremony, Sione Ika. This small, chubby, jocular man embraced a small guitar and sang enchanting songs from the South Pacific. He occasionally would stop singing, stand up, and do an unusual dance with sudden movements of the arms and legs while muttering one-syllable sounds such as "umph, ah, hoo, pi, ka," and so on. Then, he would sit back down and continue explaining the purpose of the kava ceremony.

"With dusk approaching, and the palm trees swaying in the wind, the villagers round up around the kava bowl. Kava, to us, is a symbol of the covenant. It is important to us in many ways— socially, culturally, and religiously. We relax, tell stories, feel good, and lose our cares in the approaching night. People with good voices, and not-so-good voices, share of themselves through their songs. Kava helps everyone feel part of the community and the village."

I had known about kava for many years but had long been a skeptic. Having been trained at Thomas Jefferson Medical School, a very traditional institution in Philadelphia, I had always thought that herbs were a fun distraction for many in the alternative field to play with, but had no serious role in medicine. They certainly could not compete with pharmaceutical drugs. Over the years, though, I have come to respect the power and benefits that many herbal products can provide.

I sat on the green turf while one of the Tongans offered me a cup of the brown liquid. He graciously bowed as he handed me the drink. It was quite obvious that he delighted in introducing skeptical Westerners to what South Pacific islanders have known for centuries.

I sipped from the cup, swirled the brown liquid in my mouth, and could tell it had some active ingredient since my mouth went slightly numb. As I continued sipping, Sione proceeded, "Kava is a healthy, natural way of relaxing. We don't need television. All of the villagers sit around and we tell stories. We share our thoughts and our hopes. We encourage each other. Cares and worries disappear—carried away by the warm ocean wind."

Sione was right. Studies have shown that the kava culture of the islands helps provide better social integration and a deepened sense of community (Lemert, 1976). One study even found that a community-based smoking cessation program combining kava ceremonies and group pledge was successful in helping almost everyone in the village give up their tobacco habit (Groth-Marnat, 1996).

The experience of that afternoon convinced me to look further into the kava story. The two cups I drank did help me relax. The Tongans were kind enough to offer a plastic bag of the root powder, and when I returned home, I continued experimenting with it. I offered it to many friends and family members who also reported feeling the relaxing effects of this root. In addition, I bought kava pills from a health food store and a bottle of the tincture. I started recommending kava to patients as an alternative to tranquilizers. The results were very promising.

However, I wasn't completely convinced until I came across the conclusions of a new study published in 1997 in the journal *Pharmacopsychiatry*. The title of the article was "Kava-kava extract WS 1490 versus placebo in anxiety disorders—A randomized placebo-controlled 25-week outpatient trial." Previous to this study, the longest one published was an eight-week trial. But, before I explain the results of this study, let me tell you some of the basics of this highly cherished South Pacific plant.

## What is kava?

Kava is the term used for both the plant and the beverage made from it. The beverage is prepared from the root of a shrub called the pepper plant, *Piper methysticum*, found in Polynesia, Melanesia, and Micronesia. The root is ground to a powder, which has a brownish color. The brownish powder is then mixed with water and consumed as a beverage, without being fermented.

Extracts from the root are placed in capsules and sold as kava in vitamin and retail stores.

## What's in kava?

As with any herbal medicine, a number of compounds contribute to its medicinal effects. The active compounds are concentrated in the root of the plant. Kava contains a variety of chemicals known as pyrones or kavalactones. Specific names of these kavalactones include kawain, methysticin, and yangonin (Rasmussen, 1979).

## Chemical Structure of Kawain

The water-soluble extract of kava contains different compounds than the fat-soluble extract. The central nervous system activity of the water-soluble extract was determined in mice to have mild pain-killing ability, but did not induce sleep (Jamieson, 1989). The fat-soluble extract had sleep-inducing and marked pain-killing properties. The researchers state, "The pharmacological effects of kava ingestion appear to be due to the activity of the compounds present in the fat-soluble fraction."

Many of the studies on kava used a standardized extract, called WS 1490, from a German manufacturer. The kava products you find over-the-counter contain the active ingredients.

## How will kava make me feel?

I have talked to several kava users, recommended it to patients, and taken it myself on occasion. I have discovered that not everyone reacts exactly the same way to this herb. This is because each one of us has a different biochemistry. Furthermore, different products on the market may contain different amounts of constituents. The form of kava, whether liquid, tincture, or capsule, may also make a difference on how you feel or how quickly you experience the effects. However, most of the time the effects are noticed within a half-hour.

The following are some feelings most users report:
- A state of relaxation, without being drugged
- Decreased muscle tension
- Sense of peace and contentment
- Increased sociability, especially with the right company, although higher doses may induce a state of withdrawal
- Occasionally, mild euphoria
- Greater clarity of thought, except on high doses
- Mental alertness is not affected, except on high doses
- Sleepiness with high doses

I have also talked to many users who did not feel any effects on kava, or what they felt was minimal. Side effects on high doses will be discussed later.

## A brief history

During the 1700s, Europeans discovered the beautiful South Pacific islands (which, to their surprise, had already been discovered by the natives). These Europeans noticed that the inhabitants of the islands, the Polynesians, were fond of drinking a ground-up powder of the root of the kava plant. This powder was diluted with water or coconut milk. The natives used kava not only for ceremonial purposes, but also during times when they needed to resolve conflicts among themselves.

We know for certain that the Polynesians have used kava for a long time, perhaps thousands of years. With the aid of special instruments, archaeologists have found artifacts that contained kavalactones, chemicals found in kava. The researchers state, "Thus it is now possible to link unequivocally kava drinking, a major aspect of the ceremonial culture of many Pacific societies, to the archaeological record" (Hocart, 1993).

It wasn't until 1966 that a German pharmacologist discovered that some of the active components within the kava root were pyrones, chemicals that have a structure similar to a 6-carbon sugar molecule.

The use of psychoactive plants was common in the South Pacific (Cawte, 1985). Before the Europeans introduced alcohol (their drug of choice), the Polynesians were drinking kava, the inhabitants of Melanesia used betel, and some aborigines in Australia were taking pituri. It appears that each culture found its own natural substance to reduce tension or reach an altered state of consciousness.

With the urbanization of many Polynesian islands, a shift has occurred. Those living in cities are more likely to consume alcohol and less kava, while in the villages, kava consumption is markedly higher than alcohol (Finau, 1982).

## The study that convinced me

Ten medical centers in the southern part of Germany participated in a study involving 101 patients (Volz, 1997). This was the first long-term, placebo-controlled trial investigating the safety and effectiveness of a kava extract in patients with anxiety. These patients were given one capsule three times a day of a concentrated extract called WS 1490, containing about 70 mg of kavalactones per capsule.

All 101 of these patients suffered from anxiety and tension. Many had a fear of public places (agoraphobia), social phobia, generalized anxiety disorder, and adjustment disorder with anxiety. (I'll discuss these conditions later.) Those who had significant medical or psychological problems were excluded.

The study lasted 25 weeks. A number of psychological tests were conducted before, during, and after the study, including a special test for anxiety, called the Hamilton Anxiety Scale. Furthermore, blood tests were carried out regularly. At the conclusion of the study, the following results were noted:
1. The short-term effectiveness of kava was superior to that of placebo. Most patients improved within the first two months.
2. The long-term effectiveness of kava was superior to that of placebo. In fact, the effectiveness of kava improved with time. The specific areas that improved included anxious mood, tension, fears, and insomnia. After 24 weeks, 75 percent of patients improved compared to 51 percent of those on placebo.
3. Patients tolerated kava well. Adverse reactions were rare.

Laboratory values, including red blood cell count, white blood cell count, platelet count, liver enzymes, creatinine (kidney evaluation), proteins, and glucose were not affected. Blood pressure and heart rate did not show any changes. More patients on placebo dropped out of the study than those on kava. This indicates that those on kava did not have any problems with it.

The researchers wrote, "These results support WS 1490 [kava] as a treatment alternative to tricyclic antidepressants and benzodiazepines in anxiety disorders, with proven long-term efficacy and none of the tolerance problems associated with tricyclics and benzodiazepines."

A review of the medical literature published on kava indicates a number of short-term studies that have found results consistent with this 1997 study (Kinzler, 1991).

## For what conditions is kava useful?

As a rule, ingesting a kava pill or drinking the liquid preparation can induce a tranquil state of relaxation.

Inhabitants of the South Pacific used kava for ceremonial and recreational purposes. However, kava extracts can be used for therapeutic reasons. The study by Volz published in 1997 that I just discussed has confirmed the effectiveness of kava for various forms of anxiety.

Before I discuss the use of kava, it's important to define the different ways anxiety can manifest, and the different terminology doctors use to define several subsets of this condition (Banazak, 1997).

**Adjustment disorder with anxiety**
   Anxiety caused by recent stress

**Generalized anxiety disorder**
   Constant anxiety and worry, without any particular cause

**Obsessive-compulsive disorder**
   Repetitive behaviors and intrusive thoughts

**Panic disorder**
   Sudden episodes of overwhelming anxiety

*Phobias-*

**Agoraphobia**
Fear of being trapped in a public place

**Social phobia**
Fear of social embarrassment

**Specific phobia**
Fear of a specific object or situation

**Post-traumatic stress disorder**
A traumatic event re-experienced, creating anxiety

Let's also keep in mind that several medical conditions can cause anxiety. Some of the most common are angina (chest pain), heart irregularities, and hypoglycemia. It's important to have a full medical evaluation to rule out any physical causes of anxiety before proceeding to any type of medications.

Of course, before starting medicines, one should try all other options to reduce anxiety. These include cognitive-behavioral therapies, muscle relaxation techniques, relaxation books and tapes, guided imagery, yoga, breathing exercises, meditation, sincere talks with close friends and family, and vacations.

In addition to its use in anxiety disorders, kava has been tested for the following conditions.

**Menopausal symptoms**

Menopause can be difficult for some women since a variety of hormonal changes may disturb mood and cause tension. In certain women, kava can provide temporary relief. In a double-blind, placebo-controlled study involving 20 women with menopause-related symptoms, kava extract was given at 100 mg of kavalactones per day (Warnecke, 1991). Benefits were noted within the first week. Most of the women noticed improved mood and well-being, and less anxiety.

**As a painkiller**

A study on mice demonstrated kava's mild pain-reducing properties (Jamison, 1990). A number of extracts from kava were found to be effective, including kawain, dihydrokawain, methysticin and dihydromethysticin. How these extracts work is not fully known. However, they do not appear to involve the same brain chemical system as regular painkillers, such as codeine or morphine.

When I used to work in an emergency room, I treated quite a few people for narcotic overdose. The proper treatment, in addition to inducing vomiting to get the pills out, is an intravenous injection of a medicine called naloxone. When this medicine is injected, it goes to many parts of the body, including the lungs and heart, and reverses the effects of narcotics or opiates.

In the mice study mentioned above, naloxone did not effectively reverse the pain-killing activities of kava extracts. The researchers state, "The analgesia induced by kava occurs via non-opiate pathways."

Many people are allergic to painkillers such as codeine. Perhaps we will find that kava can be useful in mild or moderately painful conditions for these codeine-allergic individuals.

Specific conditions that some patients have used kava to treat include headaches, neck pain, back pain, temporomandibular joint syndrome (TMJ), toothaches, and others. Don't expect significant relief as you would from codeine. However, kava could provide enough relief to make the discomfort tolerable.

**For insomnia**

David Snow, the host of the radio show "Doctor Health," in Honolulu, Hawaii, says, "The effects are subtle, but I sometimes sleep better with kava than melatonin. I feel super when I wake up in the morning, without the residual grogginess."

Generally, the dose used for relaxation is 40 to 80 mg. However, if you want to induce sleep, a dose of 100 mg or more may be necessary. Kava can also be combined with other herbs, such as valerian, that are known to induce sleep. Please remember that any time you combine two substances that have a sedative effect, the doses must be minimal. For instance, if you combine melatonin with kava, your dose of kava would be much less. You can take 0.3 mg of melatonin with 40 to 60 mg of kava about an hour before bed.

Jet lag is another condition for which kava may be helpful, although it doesn't address the primary problem of re-adjusting the circadian clock as well as melatonin does.

**Short-term use to reduce mild stress**
David Essel is the host of the nationally syndicated show, "David Essel Live," on Westwood One Entertainment out of Ft. Myers Beach, Florida. He says, "While I was writing a book proposal, I felt very stressed. There were deadlines to meet. Taking kava daily for a period of seven days took the edge off, and I was able to be more productive."

**Before meetings or counseling sessions**
Linda Ligon, president of Interweave Press in Loveland, Colorado, says, "Kava provides a leveling of mood. We sometimes use it at the office before meetings. It makes us kinder and gentler. It provides a subtle, uplifting effect, sometimes with a mild euphoria."

For centuries in the Pacific Islands, chiefs would use kava before and during their meetings with other leaders. Terry Willard, Ph.D., president of Wild Rose College of Natural Healing in Calgary, Alberta, Canada, says, "Our counseling department gives kava a half-hour before a session to couples who are having marital difficulties. As a consequence, there are fewer arguments between them. It's a great arbitration herb."

**For bladder infections**
Lise Alschuler, N.D., is chair of the Botanical Medicine Department at Bastyr University in Seattle, Washington. She says, "An interesting use of kava is as an analgesic (pain medicine) for bladder infections. It has a numbing effect to the bladder mucosa. I prescribe a lower dose four or five times a day. This is, of course, in addition to other medicines that fight the bladder infection."

## Other properties of kava

In laboratory studies, extracts of kava have demonstrated possible anti-convulsant activities (Gleitz, 1996). This means they would reduce the incidence of seizures.

Kavain, one of the components of kava, inhibits platelet aggregation and thus can have some influence as a blood thinner (Gleitz, 1997). The clinical significance of this property is currently not known.

## How does it work?

Since there are a number of different compounds within the kava root, it is quite likely that an extract of kava will influence a number of areas within the brain and body.

Many kava users notice that their muscles are relaxed. This is because compounds in kava go directly to the muscle tissue and reduce contraction (Singh, 1983).

A study on brain tissue showed that compounds in kava have the ability to enter the brain (Keledjian, 1988) and attach to a number of areas in brain tissue (Jussofie, 1994). The most prominent areas influenced were the hippocampus, amygdala, and medulla oblongata. The cerebellum (an area in the brain involved in coordination) was not influenced much. This is consistent with the clinical effects of kava ingestion: A regular dose of kava has little effect on coordination.

This study also determined that one of the receptors in the brain influenced by kava extracts was GABA. This stands for gamma-amino-butyric-acid. GABA receptors are influenced by a number of medicines and drugs. For instance, Valium, the best known sedative, acts on GABA receptors to make us sleepy and relaxed. Apparently, kava extracts are able to influence GABA receptors as well.

Another possible mechanism includes blocking the action of a brain chemical known as dopamine (Schelosky, 1995).

## What's the right dosage?

Everyone will respond to a different dose. However, here are some guidelines to help you get started.

Start with a daily dose between 40 and 80 mg of kavalactones. For instance, if your pill is 250 mg, and it's a 30 percent extract, then 250 mg times 30 percent equals about 75 mg. This would be adequate. Most users like taking their first kava dose in the late afternoon or evening. If one dose a day is not enough, you can increase it to two or three times a day.

As for its sleep-inducing effects, you will need a higher dose, such as 100 mg to 250 mg of kavalactones, about an hour before bed. First try a lower dose of kava and see how you feel on it before taking a higher dose. This is a prudent approach to any type of medicine or supplement.

## What you will find in your vitamin or retail store

Pills, capsules, liquids, tinctures, and teas are some of the ways kava is marketed. The tinctures are made by soaking the root material in an alcoholic liquor such as brandy or gin.

Here are some examples of products I recently found in a local health food store:
- 150 mg kava extract of 30 percent kavalactones, yielding 45 mg
- 250 mg kava extract of 31 percent kavalactones, yielding about 77 mg
- 390 mg kava, with no percentage of kavalactones listed
- 450 mg kava, with no percentage of kavalactones listed
- 500 mg kava, with no percentage of kavalactones listed
- 25 mg kava extract of 30 percent kavalactones, yielding 7.5 mg. This was marketed in a liquid formula and two teaspoons provide this amount. The liquid product also included passion flower, chamomile, and hops.

Whenever the percentage of the kavalactones is not listed, it's hard to tell how much of the active ingredients you are getting.

Tinctures and liquid may be a good way of taking kava, but, again, it's difficult to know how much of the active ingredients you are ingesting. Some people, though, are believers. One of my patients explained, "I didn't have much luck with kava in tea or capsule form, but recently I tried the tincture and so far it is the most helpful approach to anxiety that I've found." Others like the traditional form of kava in liquid form, while many find the capsules perfectly suitable.

You may not know exactly how much kavalactones you are getting in the tinctures, liquid, or tea. But, more important than knowing the exact amount of the active ingredients, is how you feel on the product. Try two or three different products and forms to see which one suits you best. As with melatonin, I have found some patients prefer the sublingual form, others like the timed-release, and still others like the regular pills. Every person has a different preference.

The amount and proportion of the different kavalactones in a capsule can depend on a number of factors, including the age of the root when harvested, the quality of the soil, the grinding process, and the concentration of the powder mix.

## How long can I take kava?

Kava can be taken occasionally as long as you wish. However, as with many medicines, daily use for more than a few months is discouraged unless it is absolutely necessary. South Pacific islanders have consumed kava on a regular basis for hundreds of years. In some Polynesians, side effects have occurred.

Daily use could lead to slight dependence. In one study on mice, high doses of the liquid extract induced tolerance within a short period of time. However, kava resin did not induce tolerance (Duffield, 1991).

The longest that kava has been given daily to humans in a well-controlled study is 25 weeks (Volz, 1997). No significant side effects were reported. However, at this point, it is still best to limit your regular intake to no longer than four months; unless, of course you are being monitored by a healthcare practitioner.

## What about side effects?

Heavy consumption of kava is associated with a skin disorder characterized by a scaly rash and eye irritation. Members of Captain Cook's South Pacific expeditions observed these skin signs more than two centuries ago.

In a 1990 study, 200 male kava drinkers in the Tonga Islands with skin changes were interviewed and examined (Ruze). All these individuals consumed kava on a daily basis. Out of the 200 drinkers, 29 had prominent skin changes. These 29 were randomized to receive either 100 mg of nicotinamide (a form of niacin) or a placebo daily for three weeks. The reason this B vitamin was chosen was because the skin lesions looked like pellagra, a severe form of niacin deficiency. After three weeks, there was no major change in the skin condition. Apparently, heavy kava consumption on a daily basis causes a scaly rash that is not due to niacin deficiency. The reason for this skin condition has still eluded science, but some researchers think kava extracts may possibly interfere with cholesterol metabolism (Norton, 1994).

Very heavy consumption of kava (40,000 mg a day of the crude root powder) results not only in a scaly rash, but harms the liver, heart, and lungs (Mathews, 1988). There have been rare reports in the medical literature of an allergic reaction to kava extract (Suss, 1996). A high dose of kava in certain individuals can lead to skin reactions and disturbances in coordination. Vision can sometimes be temporarily affected with a reduced near-point accommodation, i.e., the ability to focus on objects a few inches away (Garner, 1985).

Over consumption of any supplement, herb, nutrient, food, or hormone is not advised. This is also true of kava. This herb has medical benefits if used appropriately, but it should not be used excessively.

## Caution

Although kava does not, in the prescribed dosages, produce the mental disorientation that benzodiazepines (Valium) and other tranquilizers do, caution is still advised. **Do not use kava if:**
- You're planning to drive or operate heavy machinery, unless you are an experienced kava user, you're using a low dose, and know exactly how you react to it.
- You have Parkinson's disease. Kava could possibly block dopamine receptors, which could be counterproductive.
- You are severely depressed.
- You are pregnant or nursing. You should avoid kava unless it's recommended by a physician when other alternatives are riskier.

If you are elderly, or are medically very frail, you should be careful any time you take supplements. Always take the lowest amount. This could even mean opening a capsule and taking a third of a dose initially.

If you tolerate this low dose, the next day you can take half a dose, and continue increasing it daily. Of course, any time you have a coexisting medical condition, or are on other medicines, you should be supervised by a healthcare practitioner.

There have been rare cases of acute dystonic reactions on kava (Schelosky, 1995). A dystonic reaction is a sudden spasm of a part of the body, including the neck, trunk, or tongue. The doses reported were generally more than 100 mg of kavalactones and all these cases were reported in Europe.

Of course, as mentioned previously, kava is not intended to be used on a regular basis and at a high dose for long periods, since it can potentially lead to the above-mentioned side effects.

## Kava compared to pharmaceutical drugs

Kava has some of the same benefits as benzodiazepines (such as Valium and Xanax), but without the mental impairment. Twelve healthy volunteers were tested in a double-blind, cross-over study to compare the effects of oxazepam (Serax) and an extract of kava root at 200 mg three times a day (Munte, 1993). The volunteers on oxazepam had a slowed reaction time during a test of word recognition, while those on kava had a slight increase in the number of correct responses.

Dr. Hans-Peter Volz, from the Department of Psychiatry at Jena University in Germany, conducted the 1997 long-term study of kava use in anxiety disorders. He compares kava to tricyclic antidepressants (such as Elavil) and benzodiazepines (such as Valium). He says, "As to benzodiazepines, their effectiveness in anxiety is beyond any doubt, and the short latency needed to establish the desired anxiolytic [relaxing] effect is a particular advantage. The two major drawbacks of this class of compounds are their dependency potential and their side-effect profile. Especially when administered on a long-term basis, benzodiazepines are difficult to withdraw. In this regard, kava is non-problematic. Withdrawal symptoms or dependency developments are not known. Also, in our trial, no withdrawal symptoms occurred in the last trial phase."

"Tricyclic antidepressants [such as Elavil], also widely used to treat anxiety disorders, have major side effects, especially for higher doses, which account for a decreased compliance, and direct cardiotoxic effects which make it impossible to administer them to patients suffering from severe cardiac conduction disturbances. Such side effects are not known with respect to kava."

What about driving a car? Dr. Volz continues, "The side-effect profile of benzodiazepines, with drowsiness as the predominant symptom, leads to restrictions in driving vehicles or operating machinery. For kava-kava, the results show no detrimental influence of this compound on various psychometric parameters." However, we must note that studies have not been published giving kava to drivers and then checking their actual performance. If you're a first-time user, do not drive a car after taking kava. You should first try kava when you are at home.

Considering the factors mentioned above, it may be wise to consider the use of kava as first-line therapy in certain anxiety disorders. If you are currently on a tranquilizer, please remember that any time you try to change from a prescription medicine to an herbal product, you should do so under the guidance of a healthcare professional.

## Can I switch from tranquilizers to kava?

Let's assume you've been on a sedative or tranquilizer, like Xanax, three times a day for a month or less. While you make the transition over a period of a week, reduce the dose to twice a day, then once a day. The day you stop taking Xanax, start taking kava once a day, at a dose of 40 to 80 mg of kavalactones. If you tolerate it well, the next day you can increase the dose to twice a day, and the following day to three times daily.

Of course, if you've been on a tranquilizer for many months or years, the transition process would have to be very slow. This could take a month or two, or even more. I recommend you make this transition under medical supervision.

## Can I combine tranquilizers with kava?

This is not a good idea. A 1996 letter to the editor of the *Annals of*

*Internal Medicine* reads, "A 54-year-old man was hospitalized at our center in a lethargic and disoriented state. His medications included alprazolam [Xanax], cimetidine [Tagamet], and terazosin [Hytrin]. His vital signs and results of laboratory studies were normal. His alcohol level was negative, and a drug screen was positive for benzodiazepines. He became more alert after several hours and stated that he had been taking a natural tranquilizer called kava for the past three days, bought from a local health food store. He denied overdosing on the kava or alprazolam"(Singh).

Of course, the combination of a few drugs and the kava probably caused this problem. But it's important to keep in mind that kava exerts some action on the nervous system, such as its influence on GABA receptors, that are similar to benzodiazepines (such as Xanax). In the case of this lethargic gentleman, the problem was more likely due to Xanax since it is more sedating than kava.

## Kava and alcohol

Alcohol doesn't mix well with most sedatives and sleep medicines, and kava is no exception. Studies on mice revealed the combination of kava and alcohol to be additive (Jamieson, 1990). In other words, alcohol increased the effect of kava. The researchers state, "This interaction of kava and alcohol has important clinical and social consequences since, in contrast to traditional usage, kava is now often taken in conjunction with alcoholic drinks."

A small dose of kava combined with a small glass of wine or less than six ounces of beer should not present any significant problem. But everyone is different, and if you are sensitive to sedatives, I urge caution.

## Combining kava with other nutrients and hormones

For insomnia, you may be able to combine kava with the following:
• Valerian is a fern-like plant. Its root has compounds, such as valerenic acid and valeprotriates, that influence GABA receptors in the brain, leading to relaxation and sleep. The usual dose for sleep is

about 300 to 400 mg. When combined with kava, the dose should be reduced by half. For instance, kava can be taken at 40 mg, and valerian at 100 to 200 mg. If you are particularly sensitive to medicines, use a smaller amount of these herbs initially.

Hops, passion flower, and chamomile are other herbs you can combine with kava. As always, to be on the cautious side, initially take half a capsule an hour before bed to see how you tolerate these herbs, or half the recommended dosage.

With melatonin, it is best not to exceed a dose of 0.5 mg when you combine it with other sedative herbs.

Kava can be safely combined with vitamins and minerals.

## Interviews with kava experts

**Lise Alschuler, N.D.**, chair of the Botanical Medicine Department at Bastyr University in Seattle, Washington, regularly uses botanical medicines in her clinical practice.

*Kava is a good option for the therapy of anxiety. I prescribe a capsule three times a day to patients who have anxiety, but I prefer the tincture. The tincture is a 1-to-5 concentration and I prescribe 2.5 ml three times a day.*

*In my experience, more than half of my patients have some relief with kava. There have been no side effects because I use low dosages. Sometimes I combine the kava with lemon balm. The longest I have had a patient on kava is six months. This patient has not had side effects to the kava. However, I normally prefer to use kava for shorter periods of time.*

*As to insomnia, kava can be combined with valerian, skullcap, and wild oats.*

*In some cases, I add kava to a headache formula, especially if the headache is of muscle tension origin.*

*I would like to caution people that kava can be misused. There is a potential for toxicity on high doses if used regularly.*

*I don't see any medical reasons to combine kava with St. John's wort.*

*As to valerian root, it has more of a sedative action and is more appropriate for sleep. If a high dose is taken during the day, a person can be lethargic. Kava, in contrast, can reduce anxiety without the lethargy.*

**Terry Willard, Ph.D.**, an experienced herbalist, is president of Wild Rose College of Natural Healing in Calgary, Alberta, Canada. He started recommending kava in 1978. He occasionally uses kava himself to induce calmness and relaxation.

*The effects most people notice include a reduction of anxiety without interference in thinking. People are able to drive without problems, unless an extremely high dose is ingested.*

*Some people find that it helps with fibromyalgia. Cystitis (inflammation of the bladder) is another medical condition where it could be helpful, since kava has some analgesic properties. Kava may also have anti-fungal and antibacterial actions.*

*The effects normally last four to six hours. When I take it, I notice it within a few minutes.*

*I've been to kava parties where the discussion became deep as a consequence of people using it.*

*As far as libido, there's probably a decrease in males and an increase in females.*

*One day, I took 4,000 mg of the extract, containing 1,500 mg of kavalactones. I didn't feel any toxicity or symptoms of overdose.*

**Logan Chamberlain, Ph.D.**, is publisher of *Herbs for Health* magazine.

*I use it at the office sometimes if I'm expecting some kind of stress. Kava allows me to function without anger or pressure. A calm feeling comes on, but mental alertness stays sharp. If a large dose is ingested, I find myself withdrawing from stimulation and becoming introspective. I have taken 10 times the recommended dose without problems. I don't think kava, though, provides any psychological insights.*

**Bob Martin, D.C.**, is a chiropractic doctor and a radio show host on KFYI-AM 910 in Phoenix, Arizona.

*Kava works well for anxiety. I recommend it to be used usually toward mid to late afternoon. Patients tell me that kava makes them feel like they are on a small dose of a tranquilizer, but without the fuzziness. The thinking process is still clear.*

**Jan McBarron, M.D.**, is board-certified in Preventive Medicine and practices in Columbus, Georgia.

*I take two kava pills half an hour before going to the dentist. It helps me relax. This herb works great for anxiety. The majority of users respond to it. I believe it is a great alternative to Xanax. At least half of my patients are able to give up Xanax.*

**Rob McCaleb** is president and founder of Herb Research Foundation in Boulder, Colorado.

Sahelian: *There have been cases of kava dermopathy in the Polynesian Islands from heavy drinkers of this root. Have you heard of any skin problems associated with kava in the United States?*

McCaleb: *I attended a symposium on kava recently. None of the researchers presented any evidence that users in the United States or Europe have had this problem. Perhaps it's associated with a compound in the fresh root? We're not sure.*

Sahelian: *The kava drunk by Polynesians is the water extract of the root. What about the capsules sold in the U.S.? Would they have a different content of active ingredients?*

McCaleb: *Probably. But let's keep in mind that many fat-soluble compounds do make it in liquids. For instance, many aromatic oils in teas find themselves in the liquid. I would suspect, with our current extraction techniques, that the capsules sold in vitamin stores would have more active ingredients than just drinking the liquid.*

## Summary and practical recommendations

If you currently have a mild or moderate case of anxiety, and you need temporary help, kava may help (assuming, of course, that you've tried non-pill approaches without success). Kava is a good alternative to tranquilizers currently prescribed for anxiety. If you have a persistent case of anxiety, and kava is not effective, you may have to temporarily resort to prescription medicines.

- Start with one kava capsule of 40 to 80 mg kavalactones in the evening for three days. If this is effective, stay on this dose.
- If the evening dose is not effective by itself, take a second dose during the day. Continue on this twice-a-day dosing for another three days.
- If you still feel tense and anxious, add a third dose, and spread these three doses throughout the day.

Once you are better, you can reduce your dosage and go off the kava. If, with time, your symptoms return, you can temporarily start taking kava again.

Please remember that kava is not meant to be consumed on a daily basis for more than four months. As with any medicine, it should be used as needed, and then discontinued.

# References

**Almeida JC** and Grimsley EW: Coma from the health food store: Interaction between kava and alprazolam. *Ann Int Med* 125:940-1, 1996. Comments: There's a definite bias here in the title of this letter. Why isn't it called "Coma from the pharmacy?" The drug alprazolam is much more likely to induce coma than kava.

**Banazak D:** Anxiety disorders in elderly patients. *J American Board Family Practice.* 10:4:280-289, 1997.

**Cawte J:** Psychoactive substances of the South Seas: betel, kava, and pituri. *Asut NZ J Psychiatry* 19:83-7, 1985.

**Duffield PH** and Jamieson D: Development of tolerance to kava in mice. *Clin Exp Pharmacol Physiol* 18:571-8, 1991.

**Finau SA,** Stanhope JM and Prior IA: Kava, alcohol, and tobacco consumption among Tongans with urbanization. *Soc Sci Med* 16:35-41, 1982.

**Garner LF** and Klinger JD: Some visual effects caused by the beverage kava. *J Ethnopharmacol* 13(3):307-11, 1985.

**Gleitz J,** Friese J, Beile A, Ameri A and Peters T: Anticonvulsive action of kavain estimated from its properties on stimulated synaptosomes and sodium channel receptor sites. *Eur J Pharmacol* 315:89-97, 1996.

**Gleitz J,** Beile A, Wilkens P, Ameri A and Peters T: Antithrombotic action of the kava pyrone kavain prepared from *Piper methysticum* on human platelets. *Planta Med* 63:27-30, 1997.

**Groth-marnat G,** Leslie S and Renneker M: Tobacco control in a traditional Fijian village: indigenous methods of smoking cessation and relapse prevention. *Soc Sci Med* 43:473-7, 1996.

**Hocart CH,** Fankhauser B and Buckle DW: Chemical archaeology of kava, a potent brew. *Rapid Commun Mass Specrom* 7:219-24, 1993.

**Jamieson DD** and Duffield PH: Positive interaction of ethanol and kava resin in mice. *Clin Exp Pharmacol Physiol* 17:509-14, 1990.

**Jamieson DD** and Duffield PH: The antinociceptive actions of kava components in mice. *Clin Exp Pharmacol Physiol* 17:495-507, 1990.

**Jamieson DD,** Duffield PH, Cheng D and Duffield AM: Comparison of the central nervous system activity of the aqueous and lipid extract of kava. *Arch Int Pharmacodyn Ther* 301(Animal):66-80, 1989.

**Jussofie A,** Schmiz A and Hiemke C: Kavapyrone enriched extract from *Piper methysticum* as modulator of the GABA-binding site in different regions of rat brain. *Psychopharmacology* 116:469-74, 1994.

**Keledjian J,** Duffield P, Jamieson D, Lidgard R and Duffield A: Uptake into mouse brain of four compounds present in the psychoactive beverage kava. *J Pharm Sci* 77(12):1003-6, 1998.

**Kinzler E.** Kromer J and Lehmann E: Effect of a special kava extract in patients with anxiety, tension, and excitation states of non-psychotic genesis. Double-blind study with placebos over 4 weeks. *Arzneimittelforschung*, 41:584-8, 1991.

**Lemert EM:** Koni, kona, kava, orange-beer culture of the Cook islands. *J Stud Alcohol* 37:565-85, 1976.

**Mathews JD** and Riley MD, *et al*: Effects of the heavy usage of kava on physical health: summary of a pilot survey in an aboriginal community. *Med J Aust* 148:548-55, 1988.

**Munte TF,** Heinze HJ, Matzke M and Steitz J: Effects of oxazepam and an extract of kava roots *(Piper methysticum)* on event-related potentials in a word recognition task. *Neuropsychology* 27:46-53, 1993.

**Norton SA** and Ruze P: Kava dermopathy. *J Am Acad Dermatol* 31:89-97, 1994.

**Rasmussen AK,** Scheline RR, Solheim E and Hansel R: Metabolism of some kava pyrones in the rat. *Xenobiotica* 9:1-16, 1979.

**Ruze P:** Kava-induced dermopathy: a niacin deficiency? *The Lancet* 335:1442-5, 1990.

**Schelosky L** and Raffauf C, *et al*: Kava and dopamine antagonism. *J Neurol Neurosurg Psychiatry* 45(5):639-40, 1995. Case histories that indicate that the sedative effects of kava might result from dopamine-antagonistic properties. "This possibility is also supported by clinical findings of beneficial effects of kava on schizophrenic symptoms in Australian Aborigines" (Cawte, 1985). Experimentally, kava extracts have been shown to antagonize stereotypes induced by apomorphine in mice. We draw attention to the potential of extrapyramidal side effects of kava preparations and caution their use, particularly in elderly patients.

**Singh YN:** Effects of kava on neuromuscular transmission and muscle contractility. *J Ethnopharmacol* 7(3):267-76, 1983.

**Singh YN:** Kava: an overview. *J Ethnopharmacol* 37:13-45, 1992. Great review article.

**Suss R** and Lehman P: Hematogenous contact eczema caused by phytogenic exemplified by kava root extract. *Hauzarzt* 47:459-61, 1996.

**Volz HP** and Kieser M: Kava-kava extract WS 1490 versus placebo in anxiety disorders— A randomized placebo-controlled 25-week outpatient trial. *Pharmacopsychiat* 30:1-5, 1997.

**Warnecke G:** Psychosomatic dysfunctions in the female climacteric: clinical effectiveness and tolerance of kava extract WS 1490. *Fortschr Med* 109:119-22, 1991.

For information on kava on the internet, refer to Lee Kagan's Kava Page.

You can also see posts about kava on various newsgroups, including misc.health.alternative, rec.drugs.smart, alt.folklore.herbs, alt.support.anxiety, alt.drugs, and sci.med.nutrition.

Additionally, you can do a search for kava on the search engine dejanews, or see web sites pertaining to kava through webcrawler or altavista.

# Additional books of interest
## by Dr. Ray Sahelian

### Creatine: Nature's Muscle Builder

Is there actually a safe, natural supplement that can build muscle? Absolutely! Creatine, a combination of amino acids, has been found to be crucial for movement and muscle mass development. In this book, you will learn, in an easy-to-understand manner, how creatine works and what dosage is best for you.

Paperback, 132 pages • $11.95
Includes shipping & handling

### DHEA: A Practical Guide

This book discusses the safety of DHEA, how it affects the brain, heart, and immune system, and what is known about its anti-aging potential. More than 200,000 copies sold!

Paperback, 158 pages • $11.95
Includes shipping & handling

### Melatonin: Nature's Sleeping Pill

• Sleep like a baby
• Improve your mood
• See vivid dreams
• Prevent jet lag
• Have more energy
• And possibly live longer

Paperback, 144 pages • $15.95
Includes shipping & handling

Available at your local health food store or by calling:
**U.S. 1-800-477-2995 or 920-434-3838**
**Canada 1-888-292-2229 or 604-421-5887**

# Visit the web site www.raysahelian.com for the latest updates!

## Pregnenolone: Nature's Feel Good Hormone

Dr. Sahelian discusses the safety of pregnenolone and how it can improve vision, hearing, and mood. He looks at how pregnenolone can be used in hormone replacement therapy to help us retain our youth, as well as its role in a variety of conditions, including arthritis, depression, PMS, and various neurological disorders.

Paperback, 157 pages • $11.95
Includes shipping & handling

## St. John's Wort: Nature's Feel-Good Herb

Medical studies have shown St. John's wort to be a good alternative to prescription antidepressants, as discussed in *Newsweek* and on the popular television show *20/20*. This definitive guide discusses the practical uses of the herb, along with its safety and how it interacts with medicines and nutrients.

Paperback, 32 pages • $4.00
Includes shipping & handling

## Glucosamine: Nature's Arthritis Remedy

It's time the medical community realizes that natural nutrients have a role in the care of arthritis patients. A full discussion of glucosamine hydrochloride and sulfate, and the role of chondroitin are included.

Paperback, 27 pages • $4.00
Includes shipping & handling

Available at your local health food store or by calling:
**U.S. 1-800-477-2995 or 920-434-3838**
**Canada 1-888-292-2229 or 604-421-5887**

# No hype, just the facts

*Longevity Research Update*
Eight-page newsletter

Research in hormone replacement therapy, nutrition, and longevity is accelerating. If you wish to keep up with the latest information on melatonin, DHEA, pregnenolone, estrogen, progesterone, testosterone, growth hormone, other hormones, creatine, glucosamine, and other supplements, then this is the right newsletter for you. Dr. Sahelian and his staff constantly scan hundreds of new articles published in prestigious journals all over the world. They present a balanced interpretation of the important findings. No hype, just the facts. They also discuss advances in the field of anti-aging science and how these advances can be practically applied to improve the quality of your life.

The newsletter includes interviews with top experts, personal stories of hormone/supplement users, and a question-and-answer column. It is published in January, April, July and October for 1996 and 1997. Beginning in 1998, the newsletter will be published six times a year in January, March, May, July, September and November. Visit the web site www.raysahelian.com for the latest updates!

---

**To order by credit card, call 310-821-2409** (*Best times 9 a.m. to 5 p.m. Pacific Time, M–F*) or copy/remove this page and mail to:
Longevity Research Center, Inc.
P.O. Box 12619
Marina Del Rey, CA 90295

☐ 6 issues of *Longevity Research Update* ($21.00)
☐ 12 issues of *Longevity Research Update* ($36.00)
☐ 24 issues of *Longevity Research Update* ($60.00)

Newsletters are published quarterly (January, April, July, October) in 1996 and 1997 and bimonthly starting in 1998. Back issues of *Longevity Research Update* are available for $2.00 each. No shipping charge for mailings in the U.S. or Canada. Shipping (airmail) overseas, add 70¢ per copy.

Name _____ Phone _____

Address _____

City _____ State _____ Zip _____

Credit Card # _____ Exp. _____

We accept VISA, MC, AE, Diners Club, Carte Blanche, and JCB.

If you enjoyed reading
*Kava*, look for *CoQ10*
at your local health food store.

Paperback • $4.00
Includes shipping & handling

Available at your local health food store or by calling:
**U.S. 1-800-477-2995 or 920-434-3838**
**Canada 604-421-5887 or 1-888-292-2229**